THE FANTASTIC BOOK OF GEOLOGY JOKES: FOR EVERYONE, NOT JUST GEOLOGISTS

UNCONVENTIONAL
PUBLISHING

WWW.UNCONVENTIONALPUBLISHING.COM.AU

SHANE VAN

ISBN –978-0-6455601-2-1 PAPERBACK

DISCLAIMER: THIS BOOK IS A WORK OF FICTION AND IS NOT TO BE TAKEN SERIOUSLY; THERE IS NOTHING IN THIS BOOK THAT SHOULD BE TAKEN AS FACT, ESPECIALLY SCIENTIFIC FACT.

TABLE OF CONTENTS

Preface

Unconventional Publishing proudly presents *The Fantastic Book of Geology Jokes: For Everybody Not Just Geologists.* Nothing in this book should be taken as fact and is only here to put a smile on your face; no part of this book is meant to offend. If it does in any way offend you, you can always put it down and move on to something else.

This book is a part of a series of books that will challenge all the sciences and everyday professions. The series will cover everything from chemists to drummers. Unconventional Publishing is a company that aims to produce books that differ from the norms. Unconventional Publishing aims to satisfy society's unique and different tastes, to delve deep into the realms of science academia, and push the boundaries on what is realistically achievable, and to bring knowledge and evoke discussions into everyday life.

Geologists

What does a geologist drink?
Anything on the rocks

Want to hear some great geology jokes?
I'll have to dig deep to find them

What do geology and the Deep South have in common?
Relative dating

Geology is the most important field of the sciences, it may have its faults, but on the hole, it's very groundbreaking

Did you know that Santa is an arborist and a geologist?
Every year he has to figure out who is knotty or gneiss

My dad always gave me this advice;
"Don't be too hasty to find faults."
Good man, terrible geologist

Why are geology museums made of glass?
To keep people from throwing stones

When you are looking into The Rock's heritage, is it called geology or genealogy?

They say that a research geologist's work is always groundbreaking

A geologist works for the crater good of humanity

There was a geologist who was robbed and had his whole mineral collection stolen. He actually became the most hated man by all the local dieticians in the area. They couldn't believe he lost 100 stone in under 12 minutes

A geologist's favorite artist is Taylor Swift, they are always going around singing, "I knew you were rubble when you walked in." *I bet you sang this in your head just now*

Why are geologists so popular?
Because they make the bedrock

Who is a geologist's favorite comedian?
Chris Rock

Did you hear about the upset geologist?
He gave everyone the coal shoulder

What do geologists drink?
Anything on the rocks

A geologist dies and goes to hell and the devil takes him on a tour of the place before he finally receives his punishment.
Devil: Ok this is our last stop, see this lava lake, this is where you will be spending an eternity burning.
Geologist: Technically, we are underground, so that is magma not lava… the more you know.
Devil: You understand that this is the reason why you're here right?

Where do geologists like to eat?
The Hard Rock Café

Why do geologists binge watch Lost?
They love the island arc

How did the geologist get to work?
E-rode

Why should you never let a geologist borrow your car?
They are usually hammered and stoned

Did you hear about the drunk geologist that fell off a cliff?
They hit rock bottom

Where does a geologist sit?
In a rocking chair

Why are geologists so popular?
They are very down to earth people

Did you hear about the geologist that got divorced?
He took his wife for granite

Bob's house was absolutely rocking last night, everybody there got totally stoned. Unfortunately, his insurance doesn't cover avalanches

How do geologists greet each other?
Hi, have a gneiss day

What is a geologist's favorite band?
Rolling Stones

Why did the geologist attend Live Aid?
To get their rock on

What is a geologist's favorite dessert?
Rock-e-rode

Why do introverts make fantastic geologists?
They are always staring at the ground

A geologist's play list only includes Rock'n'Roll

A geologist is hard to date, they are always telling you about your faults

Why are geologists the best gossips?
Because they have all the dirt

Why didn't the geologist go into the haunted house?
He was too petrified

How physicists see other sciences:
Biology: squishy physics
Geology: slow physics
Computer Science: virtual physics
Psychology: people physics
Chemistry: impure physics
Math: physics minus the units

A physicist, biologist and geologist are trying to run away from fascist Germany in 1945. They are captured and then sentenced to be executed for treason to the state.
As they line up and the firing squad takes position, the physicist yells, “Lightning!” All the soldiers look up, but there is no lightning and the physicist escapes.
The biologist looks around, “um um, wolves!” he yells. All the soldiers spun around, guns ready, but there weren’t any wolves and the biologist escaped.
The geologist is stumped and starts thinking ‘boulders, no, rocks, no metaphoric rocks, heat. That’s it!’
“Fire!!!” he yells

What is the difference between a geologist and geographer?
The use of colored pencils and maths

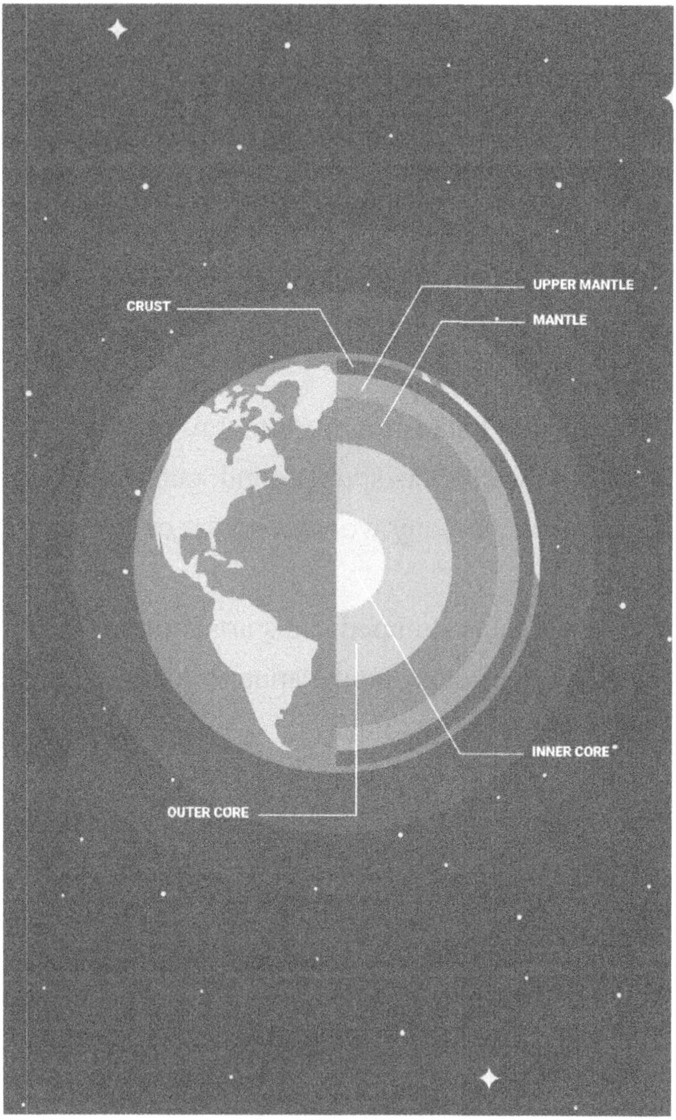
UPPER MANTLE
CRUST
MANTLE
INNER CORE
OUTER CORE

The Geo-Specialists

A geophysicist is working on his computer in his home office. The neighbors were busy renovating, and the noise was really getting his wife.
"Darling, our new neighbors are making a lot of noise. Can you take care of it?"
"Sure," he says, "what noises, I have a tonne of different filters on here. Do you know the frequency?"

Why do geophysicists like irrational partners?
They like to spend time with anomalies

A seismologist will never get a steady job

What did the doctor prescribe the seismologist?
Tec-tonic

Did you hear about the seismologist that had his equipment stolen?
It was an earth-shattering experience

When a seismologist dies, they usually request to be buried near a fault line, they are just after some good vibes

A man walks into a bar and orders 10 earthquake cocktails.
"Now that is an order of magnitude," says the barman

What happens to people who walk in a straight line during a 6.8 magnitude earthquake?
They develop Parkinson's

Geophysicists are the original hipsters, everything they deal with is underground

A man is rushing to an appointment, gets extremely lost and is now running 20 mins late.
He pulls over and asks someone for directions.
"Excuse me," he yells out at a passer-by, "I am lost and running late. Can you tell me where I am?"

"No, worries. You are sitting in a Ford Fiesta, 5 km from the town center. You have a latitude of 546^0 Northing and a longitude 783^0 Eastings. You are also sitting above large crisscrossing voids that were mined out because they showed magnetic anomalies" says the stranger.

"Wow you must be a geophysicist" says the man in the car.

"How do you know?"

"Well, you told me something that is technically correct but absolutely useless to me!" snapped the man angrily.

"Well, you must be a politician," says the man who was walking.

Smirking, the man replies, "Yes, do you recognize me?"

"No, it's because you don't know where you are or where you are going. You made a promise to a commitment without any idea of how to keep that commitment. You have also broken that commitment, but now you have pulled over and asked me for advice. I gave you technically correct advice but somehow the predicament you put yourself in is now my fault."

What did the geophysicist say to the rock when he complained about the drilling?
My sediments exactly

What did the statistician say to the geophysicist?
Your guess is as good as mine

How do geochemists eat with Potassium, Nickel, and Iron?
With a KNiFe

What is the difference between a geologist and a geochemist?
A geochemist will drink anything that is distilled, a geologist will drink anything.

Why does a geochemist think the world is so diverse?
Because it is made from alkynes of people

Why do all geochemist jokes revolve around Cobalt, Radon and Yttrium?
Because they are all CoRnY

Did you hear about the geochemist who started a rock garden?
He was so bad at gardening that four of the rocks had already died

Why was the geochemist let go?
He was an alkalic

What do you do with a dead geochemist?
You barium

What do you call a benzene ring that substituted carbon for iron?
A ferrous wheel

Geography: The Poor Man's Geology

A friend of mine once told me a story of when he was working with the Canadian Geographic Survey around the Torngat Mountains.

He had been hiking all day plotting and marking points with his GPS and had returned to the base camp starving. One of the crew got a nice warm fire started and started cooking a stew. The smell of the meat wafted through the whole camp and started making everyone's mouth water.

However, that bloody scent wafted into the forest and attracted this ginormous grizzly bear. The bear came running into the camp and charged at my friend. He didn't know what to do so he threw some shit into its eyes to blind him while he ran away.

"Where did you get the shit from?" I asked him.

"Oh, I got that from my pants."

Where are there oceans but with no water?

On a map

The best Will Smith movie is the one where he transforms into part of a map.
I am Legend

Chuck Norris once went trekking through the wilds of Siberia, when a brown bear snuck up on him. He gave the bear a roundhouse to the face. We now know this as the Tunguska event.

A geophysicist and his wife were having an argument over the phone. She was busy telling him that he was in denial.
"You need to go back to school and look at a map!! I am nowhere near Egypt!"

I heard the best geography joke the other day. You wouldn't get it, you had to be there

Geology might rock, but geography is where it is at

Thinking about the rotation of the earth, it really makes my day

How do you know the ocean is so friendly?
It waves

Where is it always a constant 90^0 but is never hot?
The North and South Poles

Two students were arguing in geography class.
"What do you think is closer, Sydney or the moon."
"The moon dumbass, you can't see Sydney from here."

What do you call a map of Alcatraz?
A con-tour map

What is white, fluffy and always knows where North is?
A Polar Bearing

Did you hear how Joe Biden plans to lower the temperature?
He wants to change from Fahrenheit to Celsius

I just realized Justin Timberlake doesn't know shit about geography.
He keeps singing about the Crimea River, it's not even a river. It's a peninsula!

Why is Bigfoot so good at reading maps?
Because he is a legend

A teacher asks her class, "Ok, what is the highest point in Japan?"
Little Jimmy straight away shoots his hand up.
"Miss, miss, I don't know that, but I know the lowest point," he yells excitedly.
"Well, what is it then?"
"Nagasaki!"

Shane Van

Plates you can't eat off

What does a tectonic plate eat for breakfast?
Quaker Oats

What does a tectonic plate eat first thing in the morning?
A continental breakfast

What did one tectonic plate say when it bumped into another plate?
Sorry, that was my fault

One tectonic plate said to the other
"Help, I am addicted to crack."
"Don't worry, it's not your fault" replied the other

Did you hear about tectonic plates who had an argument?
They had diverging opinions

What do geology and history have in common?
It's usually Europe's fault

What was the name of the punk band Pangea after they broke up?
Continental Drift

Why can't you trust tectonic plates?
Because they are all too shifty. Unfortunately, I do trust them, but I am loyal to a fault

What is the name of the next installment of the Fast and Furious movies that transverse the globe?
Continental drift

What do you call it when tectonic plates have a romantic relationship?
Subduction

What is a tectonic plate's favorite tv show?
Whose fault line is this anyway?

Why is ending a relationship like a tectonic plate?
When there is too much friction, it breaks up

The mantle and crust are having a chat when all of a sudden the crust starts shaking uncontrollably.
"Why the hell are you shaking?" asks mantle.
"Sorry, I just have a lot of pressure on my plate right now" replies crust

The Earth's surface is made up of 75% uncarbonated water. This means that, yes, indeed the earth is flat

Did you hear about the seismologist who made a new discovery?
He went mantle

How do tectonic plates greet each other?
They shake lands

Why do tectonic plates wear nappies?
Because they have in continent issues

Your momma is so fat she eats off tectonic plates

Your momma is so fat that when she does the 400m sprint, they don't measure in seconds, they measure in Richter Scale

Did you hear about the coming-of-age romance based on two tectonic plates?
The Fault in Our Lines

Why are volcanoes so pretty?
Because they are magnanimous

When I went to Hawaii I took a tour of the rim of Kilauea volcano, it really magma day

Did you hear about the two volcanoes who loved each other but couldn't get together?
They were star crossed lavas

What do you call a man with no arms, no legs, is charred and inside a volcano?
Anakin

One day, George, a seismologist was looking over the daily readings when he received a mysterious email. The email read:

George, I know who you are. I cannot disclose who I am now, but you can refer to me as 'Somebody'. I have been watching you working in your little office which others would call a tin shed and I think you are the blunt tool I need to get the message across. I have been conducting independent research in earthquake prediction and I have been successful. Unfortunately for you, George, a massive earthquake is going to occur under you in the next 36 hours. It will measure 14 on the Richter scale, it will be so huge, that your state will be broken into 2 pieces and half will get flipped into the sea. You will feel a small warning tremor in 2 hours, followed by another large earthquake measured on the Richter scale of 6 in 14 hours, but you must act now! Half the state needs to be evacuated or they will all die."

George read and re-read the email. He didn't want to start a panic, so he started calmly packing up his things, getting a van ready for departure and bringing all the necessary equipment he would need. Sure enough, exactly 2 hours later, the small little tremor was felt. George panicked and hopped in the van and started driving as fast as he could. He started calling the State Governor immediately.

"What is wrong George, I am kind of busy right now."

"You don't understand, I am in danger. You're in danger. We're ALL IN DANGER" said George manically.

"Woah, Woah, slow down, tell me what is going on."

"Somebody once told me the world was gonna roll me, and I ain't the sharpest tool in the shed."

What do you call a volcano that just watches Netflix all day?

Inactive

It is hard to get volcano home insurance in Hawaii, but the houses usually get covered.

Actually, now that the eruptions in Hawaii have settled down and are being dealt with, you can now say, "Clean-up in Isle One."
What happened to the man who opened a restaurant on the rim of a volcano?
He went bank-erupt.

What did the volcano say to the snowman?
Nice to melt you

How do you know a volcano is angry?
It blows its top

When is the best time to tell an eruption joke?
When the dust settles

What comes out of an emotional well-adjusted and sensible volcano?
Pragma

If you make a home inside a volcano, then every room is a lavatory

What happens when a volcano erupts in Greece?
Everyone eats Bak-lava

What is a volcano's favorite fruit?
Lavacado

If mountains are really just earth boners, then volcanoes must be ….

Did you know about the bird that lost all its feathers in a Pahoehoe flow?
It was molten

Did you hear about the world's shortest volcano?
They said: "Wasn't that Taal!"

Why are volcanoes the funniest land forms?
They always erupt with lav – tar

What does a baby volcano call a mommy volcano?
Mag – Ma!

Did you hear about the cult that was going to sacrifice a divorcee to a volcano?
They decided they will give it a miss

Hear about the volcanologist that cheated on his wife with his undergrad, then stole from the collection plate at church?
He fell into an active caldera, he was a real ash-hole

In Hawaii they have a marathon that runs up the side of Kilauea, it's a heated competition

I had a friend who was very out there, he would dress in bright colors, was immaculately presented, had a great skim regime and he could even float on lava.
He was very flamboyant

What kind of computer code do volcanologists use?
HTMelt

When Calbuco erupted it was hot news, however everything around it was Chile

A volcanologist was walking through a lava tube when he tripped on a rock, he Krak-a-toa
The US Geological Survey spent a lot of time studying the super volcano known as the Yellowstone Caldera. The last eruption there was 631,000 years ago with a rough eruption cycle of every 600,000 thousand years. It is now apparently overdue for an eruption. It has now been classified as having eruptible dysfunction

Fun Fact: Did you know that you can jump into a volcano? But you can only do it once

Doctor: “Hi, are you sexually active?”
Geologist: “Define active, volcanoes don’t erupt for hundreds of years and are still considered active.”
Doctor: “I’ll just write virgin.”

Joe Biden and Donald Trump are on a plane when it crashes straight into a volcano. Who survives?
The United States of America

Did you hear about the guy that committed suicide by jumping into the crater?
He suffered from severe depression

What did the volcano say to the tectonic plate?
With all this pressure you are putting on me, I am going to explode

Why are volcanoes the best rappers?
They have got a mad flow

How did a Hawaiian hipster get burned?
He walked on the lava field before it was cool

Why did the hipster visit the volcano?
To get a real lava lamp

Why is having sex with me like the St. Helens eruption?
Energetic, a big white splash, but short-lived

A politician, a priest and a schoolboy were flying on a sightseeing tour around the Hawaiian islands, when all of a sudden an ash cloud erupts in front of them. There was nothing the pilot could do except drive straight through it and watch as the ash turned to glass inside his engines and blow them up. He set the plane into a glide and went to talk to the passengers.
"I am sorry guys we are going to crash and unfortunately, we are one parachute short. I am sorry but I have a wife and children I need to take care of so I will be taking a chute" he said.
With that, he took one parachute and jumped out the plane.
The politician turned around and said "I am more important than either of you. I am the brains of this country, without me nothing would get done and people would be starving."
With that he grabbed a parachute and jumped.
The priest turned to the boy and said

"You go my son, take the last parachute. I am old, I have worked hard and done good deeds, I am ready to meet my maker."
"But father, we can both go!"
"How so?" asked the bewildered priest.
"The brains of the country took my backpack."

What did the ancient romans say when Vesuvius erupted?
At first, I was afraid, I was petrified

Why did the volcano go on a diet?
It wanted to get rid of its lava handles

There aren't too many guys in this world that would hike all the way to a volcano to throw in a ring.
But Elijah Wood

What do you find on the face of volcanoes?
A must-ash

Two geologists, Tom and Simon were having an argument on what causes more destruction. Simon was adamant that it was an earthquake. Tom said it was a volcano and Simon's argument was based on shaky grounds

Remember, volcanoes erupt to no good

If Dwayne Johnson was born in Malta, what would be his name?
Igneous, because he came from Maltan rock

With all the plastics in the oceans and the overflowing landfills, does anyone else feel like we are not using active volcanoes to their full potential?

Society has really gone haywire with its cancel culture and discrimination or what is socially acceptable over silly reasons. No one dares question when a baker has a 'bun' in the oven, but if I put 2 kids, a horse and wagon in a volcano, suddenly I am the bad guy.

Call me a conspiracy theorist, but the latest volcanic eruptions were all inside jobs

Why don't volcanoes have dinner?
They broke all their plates

Did you hear about the volcano that put on a massive fireworks display?
It was very pompeiious

Someone handed me a copy of POMPEII to watch, I refuse to watch it until I have seen POMPE, I just can't watch sequels out of order

There was a pizzeria that also sold tobacco, it was called Pompeii. Too bad the pizzas tasted like ash.

The mannequin challenge?
Pompeii, champions since 79AD

Pompeii has been the ultimate party town; everyone has been stoned since 79 AD

I went to visit the old city of Pompeii; you could say my life was in ruins

What has two eyes and use to smoke?
Pompeii

What did the cowboy say when he was exploring old Pompeii?
What in TARnation

Famous Quotes: "The floor is lava!"
Residents of Pompeii, 79 AD

This morning my kids were playing the floor is lava in the lounge room. I have been puzzling about it all day, how the hell did they get lava in there!!
They still did better than my neighbors' kids, they are so lazy they play the furniture is lava

The floor is lava is a great game, but it does have one fatal floor

How many Hawaiians does it take to change a light bulb?
None, lava lamps don't have bulbs

What do you call it when a lava is having a period?
A pyroclastic flow

What is a volcanologist's favorite spice?
Ground cumin

My partner told me that adults shouldn't play the grass is lava, but here I am sitting on the fence

Comic-Con will be held in Hawaii this year. They are hoping that any lost virgins can then be sacrificed to appease the volcano god

What does the Earth get on New Year's Eve?
A birthday quake

Your argument is so weak, it can't even be measured on the Richter Scale

I just heard the Richter Scale is no longer in use, that really shook me

The first person who witnessed an earthquake said it would have been a groundbreaking experience

What did the ground tell the earthquake?
You crack me up

The Russian earthquake made a pretty clever joke, I can appreciate its tsar-chasm

Why were the 2 earthquakes fighting?
They couldn't figure out who was at fault

What is an earthquake's favorite song?
Good Vibrations by The Beach Boys

Did you hear about the cow that got caught in an earthquake?
He made milkshakes

Which writer fantasized about earthquakes?
Shakespeare

Did you hear about the cow that fell down the fault?
She became ground beef

What did the audience say when an earthquake struck during a performance of Romeo and Juliet?
That was a true *Shakes*pearian experience

Why did all the horses run inside during an earthquake?
Because it was stable

Where is the safest place in an earthquake?
In a stationary store

Did you hear about the French cheese factory that collapsed because of the earthquake?
There was de brie everywhere

How do the Dutch know an earthquake occurred?
Their Van der Waals shook

How was the librarian killed during an earthquake?
From a title wave

Saying that… earthquake books are extremely popular, they just fly off the shelves.

Why do communists love Tsunamis?
They take the homes of the rich and let the poor have ocean front views

What is the opposite of a microwave?
A tsunami

Dying in a tsunami is Earth's wave of saying goodbye

What does an ex-wife and tsunami have in common?
It starts out wet, but it will take your house

Just after the Japanese tsunami, my Japanese girlfriend dumped me.

But as they say, there are plenty more in the sea

How did Kenobi survive the Tsunami?
He had the high ground

What is the name for cause of death by earthquake?
Death by de-fault

Knock knock
Who's there?
Knock knock?
Seriously who's there?
Knock knock knock, (sound getting louder and louder)
Ok who's there, I am coming out!!!
**dies in earthquake

Did you know that during an earthquake, coffins double as human maracas?

What is the name of an earthquake that shakes exactly 4046.86 m^2 of church land?
A massacre

What happens when an earthquake occurs on the African continent?
Djibouti shakes

Why do earthquakes make great journalists?
They have mastered the art of creating breaking news

"What earthquake?" Michael J Fox

Did you get the name of the Arabic Elvis during the earthquake?
Amal shook-up

James Bond is deep undercover in New Zealand on his latest assignment. He is fast asleep when an earthquake strikes, he is shaken not stirred. During this earthquake the local Lego museum was also affected and most of the sculptures were destroyed. Luckily enough the staff were able to pick up the pieces and move on. The local town of Taumatawhakatangihangakoauauitamateaturipuka kapikpmangahoronukipokaiwhenuakitanatahu was also hit by the earthquake, and the relief workers are still trying to figure out what the town's name was before the catastrophe. The fault caused a house's kitchen and dining room to separate, that caused a rift in the family. New Zealand doesn't have any natural occurring snakes, but the few they have in the zoo were rattled

What punishment does an earthquake get?
It was grounded

Bob and Will were watching the news and it was absolutely devasting, it was about a recent earthquake and the lives that were lost.

Bob : This is horrible.

Will: Yeah, I know

Bob: 21 dead and counting

Will: Yeah, it is horrible, counting is hard

Paying $4000 a month for a two bedroom in San Francisco is outrageous, but on the plus side, after the next earthquake, I might have a large, one bedroom apartment with ocean views

The perfect earthquake has been predicted to hit San Francisco soon, but the evidence it's based on is pretty shaky

My sister just moved to California to be a star, she just told me she felt 2 earthquakes last week. I did mention that it was her fault

If all earthquakes cease to exist on the planet, then Earth would be crack-a-lackin

What does an Etch-A-Sketch artist fear the most?
Earthquakes

Did anyone see the movie about the Mariana Trench?
It was really deep

Tsunami was feeling lonely so he decided to have a tea party. He invited Earthquake, Hurricane, Cyclone, Monsoon and Flood. No-one bothered to show up, so Tsunami had a silent T

Did you know that James Cameron suffered from depression while filming in the Mariana Trench?
He had sunk to a new low

Era

How do you turn a dinosaur into a horse?
Use an internal combustion engine

There has been a new study conducted on the extinction of dinosaurs. All the fossil evidence was collected and re-examined. It has now been agreed upon that the dinosaurs all died in car accidents. Not one seat belt has ever been unearthed with any of the fossils.

During the Pleistocene epoch, George is a stone furniture craftsman and is dragging his latest creation to Cru's who owns the store, The Furniture and Flint Cave.

George grunts, "This new chair, thick legs, very strong. Made of soft stone."

"This good chair" says Cru, "better than Bob's chair."

"Why better than Bob's chair?"

"You make legs too long, Bob cannot sit, must shorten."

Beneath their extended craniums and monobrows the two men just look at each other.

"Ok, me change."

George then drags Bob's chair back to his cave to start filing down the legs. The next day he returns to the shop and Cru sees him.

"Chair look good! Now you change Bob's desk."

"What change Bob's desk?" says a disgruntled George.

"You make too bumpy, Bob no use."

They start grimacing at each other angrily.

"Ok, me fix." grumbles George.

So, George then slides the desk all the way back to his cave to begin another night of working. The next morning, he slides the desk back to the shop. When he gets there he sees a serious look on Cru's face who standing next to Eva.

"No. No. You no do this," yells George, "I NO, SLEEP, I NO EAT, ME WANT DEER MEAT!!"

"But, changes in Eva table" says Cru

Going to a restaurant alone makes me feel like the Tyrannosaurus from Jurassic Park. Everybody is just there watching me eat.

Did you know?
Cyanobacteria was the first bacteria that evolved photosynthesis. Back eons ago, the earth's atmosphere had very little oxygen in it. The place was filled with anaerobic bacteria. Over millions of years, these cyanobacteria changed the atmosphere and caused the extinction of other bacteria. It makes me feel a little better since humans aren't the first creatures to change the atmosphere and cause mass extinction

Fossil hunters have a really unfair advantage. For starters, the fossils are already dead when they try to find them

A geologist from the Deep South could always tell the chronological order of any sandstone layer, but never their exact age. You see they were only interested in relative dating

What do you call it when someone uses 'age' instead of 'era'?
Epoch fail!

A paleontologist was trying to carbon date a dinosaur, but the results kept showing 2 different time periods. Looking at other species in each period and considering his fossil didn't have 3 anuses, he decided to rule out the Triassic period.

Did you hear about the geologist who switched careers and became a paleontologist?
He loved the rocks so much he decided to try and date them.

My ex would have made a great geologist.
She kept trying to dig up the past

When was the greatest time to tell a geology joke?
During the Stone Age

If you look carefully at some sedimentary rocks, you can see the fossil footprint of fish!

How do oil companies handle oil spills?
With slick lawyers

How long did it take to fill the Red Sea?
A very long period

Why is it so hard to date fossils?
Because they're dead

Two fossils went out to dinner one night at a hamburger joint.
The first fossil says, "Wow, this has to be the best pizza I have ever had."
"It does look really good" says the second fossil.
"Do you want to trilobite."

Why have they never found any human fossils in Iran?
Because homos do not exist in Iran

There is a new dating app that pairs up college students with old fossils, they call it carbon dating.

I was reading an article in Readers Digest about our ancestor fossil, Lucy. I had to admit that she looked kind of sexy, but no homo though.

Why shouldn't you ever believe a geologist when they say they will pay you back soon?
Because to them, a million years is relativity short

What is the name of a smell that lasts for a very, very long time?
Eon Musk.

Google is now replacing their old C++ computer code with their own new programming language. This allows all dating apps to instantly detect the age of all their users. It is because it now relies on Carbon Dating

How did Neanderthals survive the extinction of the dinosaurs?
Temporal distancing

Following the evolution of pigeons can be extremely time consuming. It took a pig eons to become one.

Why did Hollywood start investing in movies about geology?
They wanted more period films.

For those of you that think 30 seconds isn't a significant period of time. The next time your partner askes you 'if they look fat in that dress' try not saying anything for 30 seconds and see what happens.

My partner told me that period jokes are not funny at all. So, I had to throw away a whole chapter of jokes I had written on the Victorian Era.

I was asked if the era of dinosaurs actually existed. You bet Jurassic did

In the golden era before the internet, only your family and close friends knew how stupid you were.

A tourist is wandering around the halls of the Museum of Natural History when he stops and marvels at the bones of the Tyrannosaurus Rex.
"She is a beauty," says one of the guards, "that fossil is 66 million, 3 years and 6 months old."
"Wow, how can you be that precise?" asked the astounded tourist.
"That is easy, when I started here, I was told that that skeleton was 66 million years old, and I have now been working here for three and half years."

The aliens actually tried to warn the dinosaurs that they were headed for extinction, but it was all in one era and out the other

With all the toilet paper hoarding, people have to be extra careful where they eat now. You wouldn't want food poisoning, especially in this dire era.

What, or who would bomb Jurassic Park?
A dino might

Jewels that Sparkle

What did gold say to pyrite?
You're a fool and an imposter

What is a geologist's favorite magazine?
Illustrated Quartz

What is the difference between James Bond and silicon tetrahedron?
One is an iconic Bond and the other is an ionic bond

Why was the geologist confused about the new mineral he discovered?
The structure was not crystal clear

What did Darth Vader say to the geologist?
May the fornacite be with you

My grandfather wanted to be cremated and have his ashes turned into a diamond, there was a lot of pressure for it to happen

How do you make the value of a diamond drop by 80%?
Buy it

What did the gemstone say when he spilt his coffee?
Oh Garnet

Why was the geologist worried about his exam?
He was worried he wasn't going Topaz

Why don't minerals ever tell lies?
Because they are pure substances

Why did the geologist start watching boxing?
They wanted to watch the Feldspar each other

A geologist had a reoccurring issue, every morning he would come out and someone would bury his flowers in dirt, it seems the plot thickens

What do you call a fake Irish gemstone?
Sham-rock

What does a geologist call his bed?
A Mylonite

Why was coal's alibi suspicious?
It wasn't crystal clear

Why are sedimentary rocks so rude?
They like to give you the coal shoulder

Why wasn't the geologist hungry?
Because they lost their Apatite

What do you call cheese made with clay?
Terracottage cheese

Did you hear about the geologist that overdosed on clay?
He was shitting bricks

Why was the mineral confused?
They were stuck in a Corundum

Chuck Norris was walking through the deserts of Libya and Egypt barefoot. The sand doesn't burn Chuck Norris, Chuck Norris burns the sand. This was how the discovery of Impactite was made

Did you hear about what happened to the water in the geologist's cup?
It Evaporite

Two rocks decide to go out to the local bar for a drink. When they arrived, the barman greets them.
"Just the usual lads?" the barman asks.
"Yes, please," the rocks reply.
"Would you want that in pints?"
"Hell no, 2 quartz thanks."

What's a Mexican geologist's favorite day of the week?
Taconite

What is a geologist's favorite day of the week?
Carnallite

What sound does a geologist's cow make?
Moolooite

What do you call a fat geologist?
Arse-no-lite

What makes a geologist wet?
Dickite

What did the geologist do when they found Dickite?
Fukalite

Of quartz these geology puns are great

When it comes to arguing with a geologist there is no compromise, it is either ore or nothing

Why can't you build bars near ore bodies?
You can't serve alcohol to miners

Knock, knock!
Who's there?
Trick.
Trick who?
Trick ore treat!

I tried to become a miner, but I heard it was a very rocky career

What do mining trucks display on the back?
A wide Lode sign

What magazine do geologists subscribe to?
Rolling Stone

Why are rocks lazy?
Because they just lie around all day stoned

Did you see the movie The Last Jedi where Rey levitates some stones?
CGI rocks

What do you call a convict rock?
Scum of the earth

Thinking about it, the asteroid that killed the dinosaurs was technically the highest number of birds you can kill with one stone

Where does a rock eat lunch?
At the Hard Rock Cafe

Did you hear about the geologist that was trying to sell Egyptian rocks?
It was part of a pyramid scheme

What happened to the rock during its interrogation?
It cracked under pressure

Why don't boulders have any empathy?
Because they have a heart of stone

What did the rock say when he was asked to apologize?
That's just how I roll

As a geologist I am obsessed with rocks, I guess you can say that is my pro-rock-ative

A geologist was exploring the wilderness when he discovers a rock that was exactly 5280 ft long. You could say it was a milestone

What is the name of a can found in a conglomerate?
Coca-Cola-Clastic

You must be a rock and I must be moss. It must be because I'm lichen you

Did you hear about the rock that had a seizure while watching Pokeman?
It was epilithic

What do you call a rock that truants school?
A skipping stone

Where do rocks sleep?
In bedrocks

What is the name of a rock-on-rock crime?
Basalt and battery

Two boys were out exploring in the woods when they came across a naked lady having a swim. One of the boys saw her, then took off and ran as fast as he could. His friend followed.
When he finally caught up to his friend, he asked, "Why did you run away when you saw the naked lady?"
"Well, my momma always said, if you see a naked lady you will turn to stone! When I saw that lady, I felt something getting hard,"

What is a turkey's favorite rock?
Cobble Cobble Cobble

How does a rock pee?
He Dwaynes his Johnson

Which four-member rock group, doesn't sing or play any instruments?
Mount Rushmore

Why are space rocks more flavorful than normal rocks?

They are a little meteor

Why did God kill all the herbivore dinosaurs?

He wanted them to be a bit meteor

A meteorite just fell in my backyard, it came out of the blue

Why did the meteorites break up?

They couldn't comet to each other

A friend invited me to a meteor shower viewing, but I couldn't make it.

They were crushed

What do mountains like to eat?

Pom-a-granites

Knock, knock.

Who's there?

Gneiss.

Gneiss who?

Gneiss to meet you

I used to love studying metamorphisms at university, the teacher was always so gneiss

What did the metamorphic rock say during the exam?
I am under too much pressure

A few years ago, a geologist friend of mine went out into the wilderness to discover some new minerals. He found a lot of new minerals, then for about 6 months, he got stuck on his 68th discovery. Then he got to 69. Gneiss!!!

Why does everyone love The Rock?
Because he is so gneiss

Did you know that Dwayne Johnson had to lose 66 pounds for his latest spy movie. I guess you can say he is the metamorphic rock

What did the bedrock say when it was heated by a hot spot?
"Holy Schist"

What game does young aspiring geologist play?
Marbles

Geophysicists and geologists may not get along but when the schist hits the fan, they can all agree that the impact will cause the fine-grained metamorphic layers to split into irregular shapes on impact

I heard the geology department at The New Jersey University isn't what it used to be. It's been slated for removal.

My collection of metamorphic rocks was shrinking by the day. I must be losing my marbles.

What did granite say to the magmatic intrusion?
This isn't very gneiss of you

How do geologists wash?
They use soapstone

After doing a survey of Mount Rushmore, turns out it is a really good idea to put Joe Biden's face on it since it is made out of Schist

If someone is gneiss, don't take it for granite

What did granite think when sandstone said, "I am also igneous?"
That it's siliclastic

What does granite have for breakfast?
Granolar

What do you call a stone that is always complaining?
A whinstone

Why is pumice always begging for money?
It's a porous rock

There is a big crime spree at the moment. People are stealing granite kitchen tops and replacing them with fakes. You could say a lot of counterfeiting is happening

What is a rock's favorite food?
A poma-granite

If hipsters were rocks, what would they be?
Igneous, because they were magma before it was cool

What is a geologist's favorite expression?
Igneous is bliss

What did the master driller say to his apprentice?
You either do it wrong or diorite

Knock
Who's there?
Basalt
Basalt who?
Pass the basalt please

I liked carbon before it was coal

Why was the conglomerate found guilty?

They had concrete evidence

Why are sedimentary rocks so cheap?
Because they are on shale

What do you do with crude oil?
Teach it respect and manners

Why did no-one talk to the pebble?
It was rudite

Why couldn't limestone get any extra movie roles?
They were type karst

Why was the limestone so mean?
Because it was tufa

Why didn't the geologist go day drinking?
He wanted to go tillite

Did you hear about the hurricane winds that destroy the coast towns?
All the people were at a loess

Which was the most sour rock?
Limestone

Why don't geologists ever get into arguments?
They are too pelite

There was once a brilliant scientist called Dr. Holmes, who was an expert in biomechanical engineering. For years he donated all his free time and money to a special project he was working on. He wanted to create a biomechanical deer. He did succeed in creating a fawn that was part mechanical and part organic with the intelligence of a human, in fact it could 'evolve' or grow into a larger adult deer.

Dr. Holmes would come home every day and look after the fawn. After a while he actually started thinking of this young deer as his son and started talking to it like it was a human, not like a cyborg talking deer. He even taught the deer basic maths, how to read but writing was a bit of a problem.
Dr. Holmes also decided to teach this young buck the sciences as well.
The young deer was doing well with biology but started to struggle a bit with geology. There was a rock on the table which the deer couldn't identify
"Is it igneous?" asked the young fawn.
"No, unfortunately its not" replied the Doctor
"I give up, what is it then?" asked the deer.
"It is sedimentary my deer bot son."

How do you get a kidney stone?
By living a sedimentary life

Some say my humor is like a stone. Usually boring and sedimentary, until thrown at someone.

Why are geologists excellent at giving gifts?
Because they are very sedimental

What is the name of a graveyard built on conglomerate?
A sedimentary

Knock, knock.
Who's there?
Shale.
Shale who?
Shale, we dance?!

The Mountains/topography

What do you call the tributary streams that flow into the Nile?

Juve-niles

A friend of mine fell into a river in Africa, I told him he might have to check for parasites. He laughed and walked it off, I think he was in de-nile.

How do mountains see?
They peak

Did you know that rock climbers are extremely hard workers?
Do they Everest?

What is the name of a smart mountain?
Cleverest

Why do pirates always travel through mountain passes?
Scurvy

Three exploration geologists are climbing up a mountain to collect samples from the top. They are starting to scale a near vertical cliff with only a few patches of vegetation on it. As they climb, a strong gust of wind hits them and they all start falling.

The geologist in the lead grabs the branch of a small tree growing from the mountain, the second geologist just blindly reaches out and grabs the man's testicles and the third geologist manages to grab onto the shoelaces of the second guy. They all are holding on for dear life but what will be torn first?

The first geologists' vocal cords

Did you hear about the geologist that came back from the Himalayas with a gambling problem?
He liked Tibet

A geologist wanted to explore the neighboring summit, everyone else refused to go, they said it was a hard pass.

A cloud, a lake and a mountain are having an argument one day. They are all arguing about who is the strongest out of the three. They decide to settle this by seeing how many humans they can kill.

Cloud goes first and waits, cloud sits there soaking in as much humidity as it can, it grows stronger and stronger and waits until a cold front comes, then bam! Cloud releases one hell of a lightning storm, setting roofs on fire, striking people on streets.

Lake goes next, but lake has to time it, lake waits and waits, until the next full moon and the wind is heading in the right direction, then she splashes up with a storm surge and wipes out the houses closet to the water.

The mountain won by a landslide.

I finally cut the ties with a friend that was just weighing me down. Mountain climbing with a friend is hard.

Why was the geologist depressed when he got to the top of the mountain?
It was all downhill after that

What did one miner say to another miner during the landslide?
Don't quarry, I got this

A young mountaineer made a model of Mt. Everest for his school project. The teacher asked, "Is that to scale?"
"Don't be stupid, how? Just look at it!"

An exploration geologist asked a local what the weather was like on top of the mountain,
"I don't know, climate."

What do you get when you cross a desert and a mountain?
Sore and blistered feet

What is the name of the overshadowed second highest peak in the world?
Mt Everer
What does a sick mountain say?
I am feeling hill

A prospector was out in the wilds of the Amazon to look for ore veins when suddenly, a tribe of cannibals surround him. They were all just wearing loin cloths covered in red paint with long bones pierced through their noses.
"God, I'm screwed!" he says in desperation just as the clouds parted and a beam of light shined towards him.
A large powerful voice from up above was heard.
"Quick, throw your pickaxe at the tribal leader."

Like a spotlight, the light shone brightly at an old cannibal who had bright blue feathers in his hair. Without hesitation the prospector throws his pickaxe. It spins through the air and hits the tribal leader between the eyes killing him instantly. In that moment all the other cannibals start screaming and yelling, pointing their spears at the prospector.

"Ok, now you're screwed." said the voice from the heavens.

Why are mountains good listeners?
They are full of mountaineers

Did you hear about the restaurant on the peak of Mt. Everest?
It was rated 2.5 stars; the food was great but there was very little atmosphere

What do you call a volcano that never erupts?
A mountain

What do you call a kettle boiling on top of Mt. Everest?

Hy- pot-en-use

Mountains aren't just funny, they are also hill-areas

Plateaus give you the highest form of flattery

Why did the geologist take her shy friend to the quarry?
She wanted the friend to be a little boulder
Fred was an accountant, and for years he would look out of his window and wanted to be out there to explore the wilds and become a prospector.
Then one day he had enough, he quit, bought some camping equipment, a shovel, a metal detector, a sieve and off he went.
The first year was great, he found a few ounces of gold and some diamonds.
The second year wasn't too bad, he found 2 smaller nuggets and a couple of rubies.

The third year was dreadful, he didn't find anything. It got to the point where he was going to give up and be an accountant again. He was walking back home in the mountains through a snow drift, lost in his own world dreading the thought of returning to civilization when he falls down a giant ravine.
When he wakes up and looks around, he finds that he has discovered a massive ore body, and it goes for kilometers. He immediately ejaculates as he just got himself a massive ore chasm.

Technically a volcano is just a mountain with hiccups

The mountain got very little applause after his comedy set following the volcano's.
It was a tuff act to follow

Why do you have to be careful with cliff jokes?
They are too edgy

There was a geologist that wanted to go on a cruise through Norway and Iceland through all the majestic towering cliffs, waterfalls, and the glacial carved inlets. Sadly, he couldn't a fjord it

Do you know who holds the record for staying in a cave the longest?
Turns out it's a Thai

The Thai cave divers had to go through some really narrow spaces. Because of that they were given special uniforms so they wouldn't Bangkok

For a soccer team trapped in a cave, you would have thought at least one of them knew how to take a dive

What country is best known for cave paintings?
Den-mark

What kind of pants do cavers wear?
Stalactites

Did you hear about the coal mine that collapsed?

Everyone got out safely, it was just a miner inconvenience

Chuck Norris stamped his foot on the ground. We now know this hole as the Grand Canyon

What were the last words from an old geologist professor before he kicked the bucket?
"Hi class, how far do you think I can kick this bucket?" he then tripped and fell into the ravine.

Two pilots were flying over the outback near Wolfe Creek when one of them spots the crater.
"Wow, would you look at that." says the first pilot.
"It is an amazing sight isn't it." replies the second pilot.
They decide to make a few more passes over it when the steward comes up to the cabin.
"What's that?" askes the steward.
"That is Wolfe Creek, it is the site where a meteor crashed a few hundred thousand years ago and left that crater."

"That is so cool, what is that building next to it?" the flight attendant asks.
"Oh, that's the tourist center."
"Well, they got very lucky didn't they."

What do you call a duck trapped in a glacier?
A quack in the ice

What do you call someone who steals chunks of glaciers?
An iceberglar

Why don't hipsters like the Nile?
They hate anything too mainstream

Two art students set up an easel on opposite riverbanks and begin painting the scene in front of them.
One of them yells, "How do I get to the other side?"
"You are on the other side." replies the other student

Did you hear about the lunatic that jumped into a French river?
He was in Seine

Where do fish keep their money?
In a riverbank

What did the river say when the beavers started swimming upstream?
I'll be dammed

Why was the beach cove sad?
It was shingle all its life

A gorgeous woman was waving at me at the beach the other day, but there was no way I was swimming that far out to talk to her

I saw someone drowning at the beach yesterday, I threw some soap at him. They later washed ashore

What states does the Mississippi flow in?
Liquid

What has a bed, but you can't sleep in it?
A river

What happened when the river tripped?
Waterfall

I just saw a sign for Niagara Falls, turns out it was falls advertising

What did one rockpool say to the other rockpool?
Show me your mussels
Why are beaches the most confident landform?
They are 100% shore

Why was the dock's sand so resistant?
It was good at pier pressure

What do the Kardashians and the beach have in common?
They both have a bunch of plastic on them

Why did the beach get wet?
Because the seaweed

Your momma is so fat, that when she was tanning at the beach, Greenpeace pulled up with 2 buses, started throwing water at her and dragged her back in the ocean

Did you hear about how the expert of sinkholes started her career?
She fell into it.

A man comes across a very large sinkhole. He peers in and he can't see the bottom of it, so he tries to throw a pebble in and listens. He doesn't hear anything.
He then finds a bigger rock, and throws that in. He still doesn't hear anything. So then he finds a big stick and throws that it in and listens. Still no sounds are coming out. He looks around and finds a very big tree stump. He then lugs it over to the hole before he throws it in. He still hears nothing but this time a very large black dog comes running at him. He barely has time to avoid it as it leaps into the air and straight into the hole.

A few minutes later a jogger is coming past and stops the man.
"Excuse me, but have you seen my dog, he is quite large and has black fur?" asked the jogger.
"I actually have, he just ran and jumped straight into this hole." he replies.
"That can't be my dog, I left him tied to a stump." replies the jogger

With global warming, what is the worst thing about living in an igloo?
No privacy

The structure and states of the planets in our solar system:
Earth = 60% water + 30% land + 10% gases
Mars = 100 % land + 0% chocolate
Venus = 100% greenhouse effect + 0 % goddess of fertility
Pluto = 100% rock + 0 % Disney dog
Mercury = 100% rock + 0% Freddie
Uranus = 100 % gases

Did you hear about the lightning bolt that told a joke?
It was shocking

What kind of bow can't be tied?
A rainbow

Why does a hurricane have bad sight?
It only has one eye

What is a whirlwind's favorite game?
Twister

Little known fact, but Brian May, the lead guitarist of the rock band 'Queen', holds a doctorate in astrophysics. When he was writing his thesis, he angered the whole geological community and NASA. He created an odd theory about Earth's rotation which entranced a lot of people. According to him, it is 'Fat bottomed girls make the rocking world go round.'

What kind of briefs do rainclouds wear?
Thunderpants

The Dirty Geologist Relationships

What do a geologist and a prostitute have in common?
They will both hammer their subjects, and know about making the bedrock

A geophysicist asked his wife, why she married him.
"It's because you're funny."
"I thought it was because I was great in bed."
"There you go, that's my point. You are hilarious!"

What is the difference between a woman and a volcano?
A volcano never fakes an eruption

My partner loves golden meteor showers. Luckily, I have kidney stones

Why was the geologist feeling down?
His relationship was on the rocks

My girlfriend told me she wants to experiment more in the bedroom. I really don't understand why she got angry, all I did was put baking soda and a vinegar volcano on the sideboard

What does a geologist make sure you always do on a date?
That you are Cummingtonite

What is the best thing about sleeping with a geologist?
The ore-chasms

The relationship between a geophysicist and geologist was described to be like a tectonic plate. First, they rammed each other really hard, then when the friction increased, they broke up.

A volcanologist wife told her husband, "You really remind of a volcano in the bedroom."
"Is it because I explode with a big eruption and let out a loud noise?" he asked
"No, it's because most of the time you're dormant and only active around every 1000 years."

My husband and I went for a road trip through Arizona. We got into a big argument on where to go, he wanted to see the meteor crater, I wanted to see the Grand Canyon.
It was a hole thing

What happened when the two volcanoes got together?
They made hot lava

Why did the geologist couple break up?
It wasn't either of their faults but there was too much friction between them

What does a geologist do before a date?
Makes sure they smell gneiss

My wife has always had the problem of never knowing what the difference was between geography and geology. At the end of the day, she still rocks my world

Did you hear about the volcano that got a STD?
His rim was covered in shmagma

I knew a couple and the both of them were introverts. He was an archaeologist, and she was a geologist, it was just natural for both of them to start working at the museum, in fact that's where they met. Being introverts, it just made sense to have their dates at the museum. On the first date they went to the Archaeology section, on the second date they went to the geology section. The rest was history.

What do you call a party full of naked geologists having sex?
A georgy

Some geologists do have trouble with interpersonal skills, they are only used to dating rocks

A geologist's partner asked them to name all the people they had slept with in chronological order. In hindsight they should have stopped at the partners names

Why is sex with a geophysicist like an earthquake?
It lasts less than a minute and leaves you wondering what the fuck just happened.

What did the geologist say to his wife?
I lava you

Why do geologist's always get laid?
They have the best dating techniques

What is it like to have sex with a volcanologist?
Energetic, but short - lived

What do you call a hooker that pegs guys for jewelry?
Lucy in the guy for diamonds

What is wet and can spasm and shake uncontrollably?
An earthquake on a rainy day

How did the geologist describe her date?
Hematite

Why are chronic masturbators banned from the beach?
They always come in waves

Why is sex with me like having sex with a volcano?
Because later you're going to feel the burn

What did the geologist say to his wife on his wedding day?
I pumice to love you until death do us part

What did diamond say to quartz?
Nothing silly, minerals can't talc

Why is a geologist so good in bed?
They will get your rocks off
Why was the geologist student kicked out of university?
Because they were a dirty layer

Why do geologists make good comedians?
They know all the dirty jokes

Two nuns were riding push bikes down the road. One says to the other, "I've never come this way before."
"It must be the cobblestones" the friend replies

Over the last year my sexual perversions have been getting a lot more hardcore. It wasn't until I smacked a statue that I hit rock bottom

All my ex's call me sandstone. It's because I am un-dateable

Why do geologists make excellent lovers?
They spend all day looking at cleavages and studying erogenous zones
What did the paleontologist say to her boyfriend?
You definitely put the ass in Triassic

Rocky Pick-up lines

That dress looks good on you, it's marbellous

Hey babe, are you a diamond?
Because you are giving a hardness of 10

I am a geologist, so I am an expert at checking your topography

Hey babe, are you part clastic?
Because your body rocks

Hey babe, are you part clastic?
Because I want to check out your cave

Is your science major in geology?
Because I need you to check out this rock in my pants

People are 70% water, and given enough time water can pass through rock, but tonight a rock is going to pass into you

Hey babe, are you a quartz?
Because you have excellent cleavage

Hey babe, are you a sedimentary rock?
Because I love studying all the different layers of your personality

Hey babe, are you a sedimentary rock?
Because you are miles thick.

You remind me of longshore drift, every time I look at you, I get swept away, and every time you look at me, I drown in your eyes

Was your mother Medusa?
Because when you look at me, I turn rock hard within seconds

Are you a magma chamber?
Because you are giving me a hydrothermal system in my pants

Can I use your body to practice geological mapping?

Are you a rock Pokémon?
Because I really want to smash you and squirt all over you with my hydra pump

Are you the inventor of the Richter scale?
Because you rock my world

If you were to kiss the Blarney Stone, the rock would be the lucky one

Of all the rocks on all the mountains, I would pick you

Of all the rocks on all the mountains, I would pick your pair of boulders

I know my way around a map and would love to explore your longitudes and latitudes.

Are you into petrification?
Because you turned my wood rock hard

Hey, is your name silty?
Because you look so damn fine

I can't stop thinking about what is under your topsoil

Are you a group of strategically aligned atoms of carbon that follow cubic pattern?
Because you shine like a diamond

Are you made of carbon?
Because all I want to do is date you

You really rock me to my core

Are you a volcanic eruption?
Because you are on fire

When you walked in, was there a magna intrusion?
Because it suddenly got hot in here.

Are you a super volcano?
Because you just blew me away

Are you a super volcano?
Because you are very good at blowing

Are you a volcanologist?
Because you just created an eruption in my pants

What's hot and makes me hard?
Being covered in lava

Babe, are you the Archean Eon?
Because I am about to rock you

Babe, are you the Earth's crust?
Because I want to get under you, then explode on top of you

Hey girl, are you a Thai cave?
Because I want to leave some kids in you

Crust
Upper Mantle
Lower Mantle
Outer Core
Inner Core

Education

The main 2 things I learned from online dating are geography and disappointment

To be a successful prospector you don't need to have a major in geology. You just need a miner

How did the prac student drown?
Their grades were below C level

A geology student came back from Brazil. He went to his university professor and gave him a gift, a little bottle of dirt from the Amazon River. "Thank you for the sediment" replied the professor

When I was at university, I found all the geology majors used to smoke a lot of weed. I am not surprised. They were a bunch of stoners

A kid walks into his earth science class
“This class rocks!” he yells out
“That’s gneiss, too bad you are so schist at it”
says the teacher

During a field trip for the class Environmental Earth Sciences, the students were taken to a quarry that was full of sedimentary rocks. The field trip consisted of studying the different strata specimens. To be honest it wasn’t that interesting, it was just boring.

Some students were going on an excursion to Mt. Fuji. One of the kids complained saying “This blows!”

I have a double degree in geography and psychology. I wrote the leading thesis on glacial depression

What did the grad student define hydrophobic as?
A fear of utility bills.

When a geophysicist completes their doctorate, they are initiated into a knighthood. It is called the tectonic order.

A geography teacher was teaching his class how to read maps. He explained about latitude, longitude, degrees, which ones go first and where they are located on a map. He then asked his students: “What if I asked you to meet at 345 Northings and 743 Eastings?”
After a moment of silence, one brave student puts up their hand “Um I think you would be eating alone.”

My grandfather went down in history, he allegedly also got a hand job in Earth Science class, allegedly.

A geology professor was busy describing the Mariana Trench to the class. It was the deepest, coldest, most inhospitable place on the planet. His associate, Mariana walked in and told him not to talk about their date like that.

Three university graduates met up at Burger King for a quick catch up.
The engineering major was talking,
"Did you hear about the new wind turbines they are placing up in the hills? I had to work out the stress loads on the bearings of those turbines."
"Oh really?" says the geologist major, "I had to look at the underlaying rock strata and decide whether or not it was able to hold the load of the towers."
Meanwhile the art major said, "Do you want fries with that?"

A student was struggling with a mapping exercise.
"I will never get this!" grumbles the student.
"Not with that latitude" answers the geography teacher.

Glossary

A

Acid Rain: the type of rain that burns skin and doesn't get you high

Absorb: what a sponge does, think of a 2-year old's brain that mimics everything. Like when you accidentally swear in front of them, then they repeat the words over and over again

Agassiz, Luis: considered a founding father of American science because he served as a reagent of the Smithsonian Institution but was actually educated in Switzerland and France

Agricola, Georgius: in 1556 he wrote down the most influential work on mining at the time. This work included occupational disease. Mentioned nothing about sending 12 years old into coal mines in England

Agronomist: soil doctor

Albedo: the amount of radiation the earth sends back, this has increased thanks to people wanting concrete lawns

American Geophysical Union AGU: originally part of the National Academy of Sciences. But thought they could promote Earth Science better by having 2 annual meetings

Andean margin: When Andean draws a border in his workbook

Andesite: the rock that is in-between granite and basalt and contains quartz. Named because it was found at Andes' site

Aquatic: water stuff

Aquifer: a rock that pees water

Asthenosphere: the Greek goddess Asthenos area of influence or her devotees

Atmosphere: atom's balls

B

Basalt: the volcanic rock that *DOES NOT* contain quartz. Why do some junior geologists find this so hard to understand? Seriously, just look at the mineral percentage diagram. I do not understand why this is such a difficult concept to understand

BCE: it was originally called BC before political correctness got in the way. *Fyi BCE and BC have the same identical numbering system*

Biogeochemical: a reaction that couldn't make up what branch of science it should belong to

Bjerknes, Vilhem: Norwegian meteorologist that studied weather and helped in understanding circulating air mass, fronts and circulation. So, a viking was concerned with the weather? As if that never happened in the past with all their sailing and myths about fogs

Boundary: a line or limit that divides one area or zone from another. For example, the friend zone

C

Calcareous: a rock that drank lots of milk

Cellulose: fat thighs

Chlorofluorocarbons: first chemical to get a worldwide band. Great for cooling fridges though

Circulation: movements in a system, the air created by solar heat, the mantle from convention. People's genitals swelling when they see a geologist

Cleavage: the gap between a rock's boobies

Climate: Cli's friend

Condensation: how a plastic bag in the desert saves you from dehydration

Conductivity: what train drivers do

Continental crust: the crust on an Italian pizza

Convergent boundaries - what happens when 2 plates love each other very much and erupt on each other

Core: a shockingly terrible 2003 movie that had very little scientific fact, but the giant geode was cool

Coriolis effect: if movies have taught me anything, snipers use this knowledge all the time

Crust: edge of a pizza

Crystal: a solid mineral that has the same repeating atomic structure forming a lattice structure. It does not contain superpowers. It cannot recharge in the light of a full moon

Crystallization: last process when making drugs

Current: a type of fruit often confused with a grape or berry

D

Denitrification: removing of hookers (night rate) from an urban area

Diorite: Andesite but with tiny crystals. A black and white rock nicknamed the salt and pepper rock. Warning does not taste like salt and pepper and will crack teeth

Dissolved solid: what gangsters do to rivals when putting them in a drum of acid

E

Earthquake: the maps of the earth in the video game Quake

El Nino: a ~~2~~ ~~3~~ ~~7~~ random year climatic cycle in the tropical pacific when the trade winds die down

Erosion: opposite but also the same as weathering

Eutrophic: how to drown aquatic animals in a water system

Eutrophication: same as above but caught in the act

Extrapolate: the only thing theoretical geophysicists actually do

Extrusion: when something is poking out, like a bone out of the skin

F

Ferrel cell: a latitude with a wild wind

Ferrel, William : an American meteorologist & school teacher that made a tide prediction machine, also had wild parties. People found out that his parties were mostly made of hot air, so the Ferrel cell was named after him

Floodplain: for some reason people still buy and build in these locations, then are amazed when every 10 years their house goes under water

Fluorescence: glowy stuff, usually poisonous. A friend of mine once drank this and then he was able to predict the future. He said he was going to die, then he did

Fossil: my old geology professor

Fossil record: the grade my old geology professor would put on my transcript

G

Gabbro: Gabby's older meaner brother. Also the better, pretty half of basalt

Geocentric: An eccentric geologist

Geographic Information System: a fancy name for a mapping app

Geological Society of America: much more grounded than the AGU

Geosphere: a geologist's ball

Gilbert, Grove Karl: American geologist who founded geomorphology, and published on erosion and river incision. Basically, he got paid to go canoeing all day long

Glacial Period: when it's 'that time' for a glacier

GPS: depending on which company you have your phone with, but it is either an accurate or inaccurate location navigator

Granite: the rock is full of quartz, unlike basalt. This is also the pretty version of rhyolite

Greenhouse gas: farting in a greenhouse

Ground truth: when a geologist speaks without lying

Ground water: an earthy tasting water.

Gutenberg, Beno: German geophysicist who determined the depth to the core, described its elastic properties, also described differences in crusts and created a magnitude scale for earthquakes. Basically, extended and continued all of Mohorovicic's work, also loved sushi so much they named the Bento box after him

H

Hadley Cell: In which part of the jail Hadley is sleeping

Hadley, George: first started describing winds around the equator, allegedly got a prison cell named after him.

Heese, Harry: looked at the work of Marie Tharpe and Bruce Heezen, then though hmmm, I have my own idea about this. Then helped develop the theory of plate tectonics

Hindcast: the same as forecast but looking backwards in time before weather anomalies were recorded

Hot spot: Cindy Crawford's mole

Hot spot: where you can get Wifi

Hot spot: the most popular and current trendy dining establishment

Hurricane: what people yell when Cane is about to score a touchdown. "Hurry Cane!!"

Hurricane: Little known fact… but hurricanes were originally named after unpopular politicians, they blow a lot of hot air and leave a lot of damage in their wake

Hutton, James: the father of geology, got in a lot of trouble with the church because he was convinced the earth was older than what the bible tells us

Hydrosphere: a ball made up of water

I

Ice core: what is in the center of my ex's chest

Ice patch: what was in between her legs but warmed up when Johnny, the football player was around

Iceberg: a type of lettuce

Iceberg: an item that is used to sink ships

Igneous: any way a geologist describes himself, "I genius"

Immiscible: a homophobic liquid

Insulation: used to keep us cool or the power of the sun

Interglacial period: when two very close friend glaciers have their flows in sync

Intergovernmental Panel on Climate Change: a group of people that fly around the world in private jets, stay at the best resorts and function centres, eat the rarest and most expensive foods, and want to tell ME about my energy consumption

Intertropical Convergence Zone: when a mommy tropical zone and daddy tropical zone love each other very much

L

L- wave: the L stands for lots of damage

Landscape: when the mountain has a quick shave before a date

Lawes, John Bennet: a farmer who loves testing fertilizers with his research station still growing. Terrorists who love nitrogen fertilizer watch these experiments very closely

Leaching: when a 34 yr. old grown man still lives with his parents and refuses to get a job

Lenticular: Leni's nickname for his single testicle

Limestone: a stone that is crushed and used in drinks

Limiting factor: money

Limiting nutrients: running out of alcohol on boxing day

Limnologist: an expert in mixing different alcohols and limes

Logbook: a book which contains pictures of an arborist's favorite trees

Lovelock James: inventor that came up with an electron capture detector which was used to provide information on halogen-bearing chemicals and CFC's in the atmosphere. In short, his invention saved the world. His last name describes how he felt about the world, he hypothesized Gaia and he practiced what he preached and even refused to injure rabbits when inventing a heat shield, instead he tested the shield on himself. He described it as "exquisitely painful." He was also a conscientious objector in WW2, until he heard about the Nazi atrocities, then he tried to enlist. A man with a huge heart.

Luster: the more you spend on an engagement ring the more luster it should have

Lyell, Charles: championed for uniformization, wrote Elements of Geology, The Antiquity of Man which supported Darwin's theory. Another geologist who had to argue with the common establishment even though he was right

M

Macronutrient: big food, like a 72oz steak

Magellan, Ferdinand: an explorer who loved sailing everywhere and naming shit after himself. Sails here, "oh look a new penguin, I will call it the Magellan penguin," sails there, "Oh, look, no one has sailed here before, this will be the Magellan Strait," sails over there, "Oh look, new stars, they will now be called the Magellan cluster." He even opened a documentary streaming service called Magellan TV

Magnetism: the natural aura all geologists have

Mantle: a loose sleeveless cloak or shawl, worn especially by people at a renaissance fair

Matthews, Drummond: dragged magnets across the floor of the Indian Ocean and found out that the polarities were different the further away from the ridge. Then thought, 'Hey maybe Wagner was right!'

Maximum depth: really? Do you honestly need me to tell you what this means?

Mesosphere: the sphere of mess created by a teenager in their room

Metamorphism: when something changes slowly over time. E. g.: When you get a partner and become really content, then one day you look in the mirror and you are nearly 100kgs heavier than what you were

Meteorology: the study of looking for lost meteors

Microfossil: ancient bacteria

Micronutrient: a small feed that leaves you hungry. e.g., celery

Microplastic: same as above but by complete accident

Mid Ocean Ridge: every ocean has one, it is literally the belly button of the ocean

Middle Ages: with our technology, it's about 50 years old but back in the day it was 20

Millibar: a bar that only sells small drinks

Mineral: to be honest, people really only care about the expensive ones like diamonds.

Miscible: easy to miss

Mohorovicic Andrija: the Nikola Tesla of the seismic wave world. Found the difference between seismic waves and figured out a lot of the structure of the earth

Mohorovicic Line: Moho named the place where solid crust meets mantle after himself

Mohrs, Friedrich: he got hired to catalogue a private collection of minerals, instead of analyzing the chemicals which took time, he decided just to see how tough they were and scratched everything against each other, the harder it was to scratch the more money it was worth and the bigger of an erection he got. We still use Mohr's hardness today

N

Natural Resources: um, everything really, when you think about it

NCAR: Like NASCAR but without the ass

Network: how you really get jobs and promotions

Nitrogen Fixation: addicted to nitrogen

NOAA: like NASA but for the oceans

Nutrient: if you're a POW then a cockroach is one

Nutrient influx: Christmas lunch

O

Oceanic crust: pizza crust coated with sea salt

Oceanographer: someone who produces erotic photos of the ocean

Osborn, Henry Fairfield: Henry was the original "Dr. Alan Grant" from Jurassic Park, this geologist / paleologist discovered both the T-rex and Raptor

Ozone: ozone needs to be very grateful to Lovelock

P

P-wave: the smaller baby seismic wave of the group

Paleomagnetic Reversals: when the natural magnetism of a paleologist reverses and only attracts members of sex that they are not attracted to

Pangea: the super-continent that is known for supplying the best bacon also know as pancetta

Peridotite: a kryptonite look-a-like

Phytoplankton: photogenic plankton

Plate Boundary: the area that limits your intake of food at a buffet

Polar Cell: a jail cell with its thermostat turned off

Precipitation: it's all about moderation, a little is great for the garden, a lot will wash away your house and drown your dog

Prevailing Wind: the smelliest fart in the room

Q

Quartz: one of the most common minerals, hippies like to wear it around their neck thinking it gives them magical healing powers

R

Rayleigh surface wave: named after an extremely obese person. The surface wave is what you see when you slap Rayleigh's (or an obese person's) belly

Richardson, Lewis Fry: was the first person to use maths to predict the weather. He must have subtracted the one instead of adding, because no weatherman has accurately predicted the weather since

Riparian: the name of a shank in the Aryan brotherhood

S

S-Wave: second wave from an earthquake, still not the deadliest wave

Salt: it wasn't a good currency, it would dissolve in the rain

Sea: ice: what the Titanic didn't do, it didn't see the ice

Seafloor: the last thing you see when you pass out drunk

Seafloor spreading: when you pass out drunk spread-eagled

Seismic Wave: when the earth wants your attention, it gives a little wave

Seismologist: a scientist that gets really, really excited about vibrations

Serpentine: the nature of my ex, she was very serpentine

Smog: the name of a dragon from The Hobbit

Soil: what babies do in their pants

Sonar: what bats have, and what people who wake up in the middle of the night use to go to the toilet, wish they had. The brain says, "I don't need to turn the light on, I know my way around." Meanwhile the toes say, "Seriously dude, not this again, just turn the fucken light on."

Spectrogram: measurements used to weigh ghosts

Steno, Nicolaus: Danish geologist and anatomist, fell in love and wanted to present his fiancé with the perfect crystal, he researched thousands of crystals, never found it but did come up with the Law of Interfacial Angles

Stratosphere: when someone tries to mug you, kick them so hard that their nuts end up in the stratosphere

Stratosphere: The name Strato the Greek gives his single testicle

Subduction: trying to get into someone's pants

Subduction zones: when giving someone a massage, these are the zones you gently caress to give them shivers and get them excited

T

Terrestrial: being born of earth, apparently lizardfolk are terrestrial and evolved on earth before humans

Tharp, Marie: Should be considered the mother of geology, she was not allowed to work on ships at the time, so Hess sent the all the data to her. She was the one that mapped the sea floor. She also suggested returning to the theory of continental drift. She got laughed at and warned that she would end up like Wegner but continued with her argument. She didn't get the credit she deserved

Thermosphere: what you carry your coffee in

Trachyte: the mark left behind when you remove soil from your pants

Transform Boundary: The wall on Cybertron between the Autobots and Decepticons

Transpiration: what happens when you need to fart on a first date and just hold it in

Trench: Mr. T's wrench

Troposphere: the tropical rainforest area in Australia where people go crazy thanks to the heat and humidity

U

Uniformitarianism: the present is the key to the past. This only applies to geology you can't apply this theory to the stock exchange, unfortunately. Like all the best geology theories, this was not accepted at first and instead, laughed at

W

Wave action: what happens at a sporting event when everyone quickly throws their hands up and down

Weathering: the process of thinking about a choice or comparing 2 choices

Wegener, Alfred: the man, the legend. Proposed continental drift with evidence but was laughed at by the science community and labeled a conspiracy theorist because he went against science at the time. He is the reason why they say geologists are the toughest and coolest of all the scientists. Travelled all over Greenland and the Arctic searching for evidence, rescued his team after they got lost, fought in WW1 and got shot, twice. Then on his final Arctic expedition when his men were stranded, he once again hiked to get help. He was never seen again, never knowing that his "crazy" conspiracy theory was going to be proven true

Wilson, J. Tuzo: a party animal of a geologist. Wrote the hypothesis of hotspots. The formula to all the hottest and trendiest bars to drink at. If I ever open a bar, I will call it 'The Hotspot'. Tuzo is not as cool as Wegener

World Meteorological Organization: a society of like-minded individuals, who meet in secret wearing purple robes. They bring their pet meteorites and talk about meteors the whole time

Y

Yield Point: The moment a geologist gets to the airport and does a bag check-in, only to find the bag is about 10kgs overweight due to the rock specimens collected. If you go to Auckland International Airport, the terminal doors are surrounded by rare specimens thanks to the annual Macquarie University Geology Field trip.

James Hutton misquote: "As thank you for your purchase, Unconventional Publishing is gifting their readers a free joke book full of adult jokes similar to the jokes found here."

Simply visit :

https://www.unconventionalpublishing.com.au/

Other Great Books in the series!

Author's Note

Thank you for taking the time to read this book. I studied geology many years ago and worked in exploration for a while. I do love geology and all aspects of geology. I have been itching to write this book for a long time, in fact this is the longest joke book yet, with my editor yelling at me to stop.

I hope you had just as many laughs as I did in writing this book. If you did enjoy reading this, please leave a review that would be greatly appreciated, and would help me out immensely.

****Free Adult Jokes Book *****

You can also find another free joke book on the company's website, one that is not for sale, one that is not for the faint-hearted, one that can be considered a bit risqué and politically incorrect. Consider this a gift for taking the time to purchase and read my book but be warned, only download it if you are not easily offended.

www.unconventionalpublishing.com.au

If you want to see any other professions being roasted or want a particular joke on a shirt, please visit our website and let us know.

Kind Regards

Shane Van

Made in United States
North Haven, CT
06 January 2024

47125697R00085